AN ANGLER'S ALMANAC

MY FIRST DAY FISHING

Written by
WILL MILLARD

Illustrated by
JOANNA LISOWIEC

MAGIC CAT 🐾 PUBLISHING

For Grace and Macsen;
always believe in water's magic
—W.M.

CONTENTS

12-13 Tackling Up
14-15 My Very First Fish
16-17 The Earliest Anglers
18-21 Nets
22-23 Traps
24-27 Your First Rod
28-29 Freelining
30-31 Your Tackle Box
32-33 Float Rigs
34-35 Hooks
36-37 Knots
38-39 Plummets
40-43 Shotting Up
44-45 Other Tactics
46-47 Predator Fishing
48-49 Fly Fishing
50-53 Baiting Up
54-55 Essential Extras

56-57 Keeping Out of Trouble
58-59 Think Like a Fish
60-61 Watercraft
62-63 Casting Out
64-65 Waiting
66-67 Striking a Bite
68-69 Fish Safety
70-71 Capturing the Moment
72-73 Different Fish & Where to Find Them
74-75 Micro Fish
76-79 Bigger Fish
80-81 Fly-caught Fish
82-83 Ponds
84-85 Slow-flowing Waterways
86-87 Flowing Rivers
88-89 Welcome to the Club
90-91 Glossary

FOREWORD
It's almost like a magic trick.

You put a piece of bread on the end of your line and flick it into the water – and somehow, when you pull it out, it's turned into a fish!

But when I first went fishing I couldn't get the trick to work. I told myself I'd just been unlucky, and tried again. But I still couldn't catch anything. And there was nobody in my family, or anyone else, who could tell me the secret spell – or whatever it was that I needed to know. So I gave up fishing. I was six or seven years old.

But a year or two later I heard that one of my school friends knew how to catch fish. Like Will, the author of this book, he had been taught by his grandfather. And one day my friend took me fishing, to a place on the river a little way outside the village. The current was slow and the water was dark, and at a certain point my float bobbed and then went under ... and I brought in a bright silver fish with red fins. It was a roach just a few inches long, and something about that moment made me want to see what else was down there under the water. And that's what I'm still doing, more than fifty years later.

So what is the secret to catching fish? Part of it is knowing that fishing is not as complicated as it can seem. And a big part of that is getting some good, simple guidance when you start off. Best of all is having somebody in your family who can show you what to do.

But if you don't have this (or even if you do) you will find this book a wonderful source of knowledge and inspiration, as you plan and make your first casts. It will help to give you the understanding and confidence you need – and a head start to your own exploration of the underwater world. And, like me, you might be surprised where that leads...

Jeremy Wade

Presenter of *River Monsters*

THE WATER IS MAGIC

I can still close my eyes and remember everything about my first day fishing.

Especially that little fish I caught! Fishing has been there for me through all life's ups and downs. From exploring in remote jungles, to presenting television programmes and raising a family, it has been the one constant thread; but the most important thing to me now is to pass on the thrill of fishing to the next generation of anglers and get them to care about the water as much as I do.

There are many threats to fish and their ecosystems today, but I still believe that the water is magic. Even after thirty-five years, fishing keeps on bringing me right back to that young lad, who's thrilled to just catch his first fish with his grandad. And I can still be shocked too.

In writing this book, I took to fishing on a very remote loch in the Highlands of Scotland. I was working on another book alongside this one, and the person I was writing about lived alone, in a log cabin hidden right next to the water.

That loch once contained a type of giant brown trout called 'ferox', but no one had caught one there in many years. A horrible disease had gotten into the water and wiped out all of the ferox, and every big brown trout too. Today, all that was really left to catch were small brown trout and the pike that ate them.

As an angler, I was very happy to catch the little trout and try for a pike, and for several months that was all I did – with very little success to speak of. Then, late one evening, after a whole day without a single bite, I lost a fish that I knew was a big one.

I felt a huge thump on my rod as it had attacked my spinner, but after just a few seconds it spat out my hook. I assumed my chance was gone. Nevertheless, I put on a big lure intended for a large pike and tried again.

Three casts in, and I felt the almightiest crunch on my rod tip. The whole rod hooped right over and I had to run up a sandbank to keep in contact with the extraordinary creature fighting hard on the end of my line. It ran up into the mouth of a waterfall and in I went, right after it.

Cold water filled my boots as I cradled that true miracle of a fish. A 27-inch fish-of-a-lifetime – and the largest ferox to be caught from that loch in three decades. It was a fish that shouldn't really have even existed, yet there it definitely was, in my hands.

As I released the fish back into the water, I felt like the luckiest angler in the world, but one thing that angling teaches us about fishing – and life – is that no matter what, we should never ever give up; either on the fish or the places where we really hope to catch them.

Miracles can and will always happen, as long as you never stop believing that the water really is magic.

— Will

TACKLING UP

I am going to let you into a secret...

I own more rods and fishing tackle than I could ever use in my life.

I've got rods that **stretch** on for several metres, and rods that **pack down** into a pencil case. I've got **bendy** rods that were made long before I was born, and **brand-new** rods that are arrow-straight and a bit scary to cast.

And absolutely none of that means I am a very good angler. At all.

You honestly don't need to have loads of kit to get started. In fact, I think the **best anglers** are the ones who have got really good at fishing with just one rod.

In this section, I'll show you how to tackle up a **float rod** from scratch; but maybe I should start by telling you about the very first fish I ever caught. And I did that without using any rod at all...

MY VERY FIRST FISH

It is one of my first memories...

I was running down a steep slope towards a stream. Mum was shouting "Slow down, Will!" But it was already too late. My body flew forward and I fell head-first into the stream. We all laughed.

"Oh Will," said my mum. "Your boots are *full* of water!" She pulled them off one by one, and, as she poured out the water, a small, dark fish landed in front of me. Instinctively, I caught it.

That was the moment. Cupped in my hands was my first ever fish.

It was a miller's thumb and it was absolutely *beautiful*.

It flicked a spiny dorsal fin into the air and flared its gills. It was so small, but the message was clear: I'm angry and I want to go home.

Gently, I placed the miller's thumb back into the stream, and with one solid thrust of its tail it was gone. *Excitement* pulsed through my body like an electrical current. I wanted to catch more fish and I was desperate to learn how.

So, while you'll need more than a welly to catch a fish with any regularity, my first lesson was that if you can manage to do it with just a welly, then you don't need a mountain of fishing gear to be successful! If you fish well enough, a fish shouldn't notice your fishing kit, or even know you exist at all. . .

That is, until you've *hooked* it!

Miller's thumb

THE EARLIEST ANGLERS

I am going to assume that you, like me, will mostly be fishing for fun.

I'm guessing that your life doesn't depend on what you catch! That being the case, in the United Kingdom there are strict rules about what fish you can take for food, and freshwater fishing is usually strictly '*catch*' and then '*release*'.

But when we cast a line into the water, we are casting back to our ancestors, to a time when fishing was all about survival. Deep in a cave in the Southeast-Asian nation of East Timor, evidence has been found that proves **ancient people** were using fishing tackle more than 40,000 years ago, when humans were only just establishing populations across the world.

USING YOUR HANDS

The simplest form of fishing uses no equipment at all. Fishing with your hands is incredibly difficult but, in fact, many people still practise it today in the southern states of the US where fishermen use a technique called '*noodling*'.

The noodler places their *bare hands*, or a foot, into a catfish's hole in the bank...

This provokes the catfish to *bite*! The noodler then holds firmly onto the catfish's jaw and pulls it out of the water.

DO NOT TRY THIS!
Catfish are strong.
Their holes can be deep underwater and you might encounter an alligator or a snake instead!

NETS

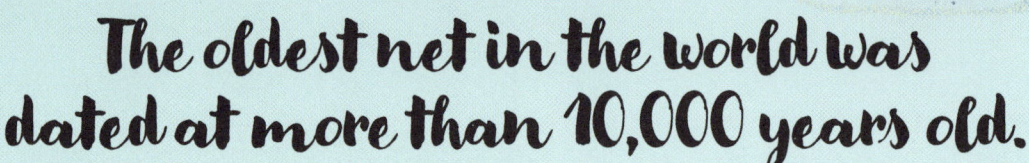

The oldest net in the world was dated at more than 10,000 years old.

It was discovered in Finland and made from strands of willow twisted into a *net lattice*, with fist-sized rocks attached to sink it into the lake. Nets were probably being used to catch fish in even earlier times; many historians agree they were likely the next step in fishing after spears, hooks, lines and hands.

WHAT YOU NEED TO POND DIP

A *net* with a rectangular-framed head, which gives you a much wider scoop area to catch bottom-dwellers.

A *magnifying glass* helps you to inspect your mini-beasts. Try drawing what you see – you'll be amazed at what you notice.

A *shallow tray* filled with water allows you to observe your creatures. The best trays are white, so you can really see what you have caught.

A *guidebook* tells you what most of the common species are.

HOW TO POND DIP

The area most likely to hold critters is right down the edge of the *bank*. Carefully run the net along this edge, as tight as you can to any plant roots. Be wary of stirring up the silt on the bottom and filling your net with mud!

Rotate net in a figure of 8

Avoid silt

Work net tight to the plants

HOW TO STREAM DIP

Position your net facing into the flow, but just below the place you want to collect from, like a large *rock* or patch of *weed*. Then, slowly lift the rock or weed and gently sweep upstream into the area you have just exposed. Afterwards, make sure to replace the rock or weed as you found it.

Flow — Rock or weed

Flow — Sweep

HOW TO CARE FOR YOUR CATCH

Once you've made a sweep with your net, quickly but carefully turn the net inside out in your *tray*.

Wait and watch. At first you might see nothing, but all it usually takes is some time to spot something.

Don't poke or squeeze the creatures you have caught: they are likely to be small and *really fragile*.

The moment you've finished looking *return everything to the water* – especially on a hot day.

Keeping safe, lower the tray below the water's surface and then gently let everything go.

Finally, check the tray and net for stowaways, then *wash* them (preferably in hot water from the tap) and let them dry out completely to kill any invisible pollutants or diseases ahead of your next expedition.

TRAPS

Fishing traps have been used since the Stone Age.

Woven baskets, fences, bags and pots were all used by ancient peoples and were – and still are – useful because they caught fish without the fisherman needing to be there at all. In the Fens, where I grew up, a traditional eel-trapper called Ernie James would work his traps, called *grigs*, in the rivers and creeks. He made the grigs himself using stripped branches from locally grown willow trees.

Eel

HOW TO MAKE A BOTTLE TRAP

1. Take a *plastic bottle*, punch in some small holes and cut off the top. Get an adult to help (you'll be a better angler if you hang on to your fingers).

2. Place your *bait* (bread, corn or meat) inside, along with a *stone* to weigh it down.

3. Push the top of the bottle back inside itself to make your *funnel*. Seal the edge with bulldog clips so that it's easy to open back up.

4. Place the bottle in a shallow area of a pond, or a flowing stream, and leave for around an *hour*.

5. Open the trap and place your catch in a jar to *briefly observe* it before returning it to the water.

YOUR FIRST ROD

When I was starting out, I wasn't too fussy about what I would use for a rod...

But I didn't catch much! People who don't know about fishing think that any old rod will do, and before you know it, you are down the local pond trying to catch a tiny roach with a broomstick designed for shark fishing in the 1980s.

Sometimes, the difference between catching and not catching can be a really small change, like using *thinner line* or a *smaller hook*. This is where this book can really help. So, let's start by looking at some basic rods that you can definitely catch fish with.

Hook

WHIPS

A whip is a long pole with a bendy tip. It is either made up of a couple of joints you can easily pull apart, or it is telescopic, for collapsing down. It tapers from the thicker end you hold, called the *butt*, right up to the thinnest end at the whip's *tip*, where you attach your tackle using a connector.

Fishing with a whip gives you a much better chance of *landing a fish* than using a stick or a beanpole, and you don't have to worry about the mechanics of a reel or casting perfectly: you just swing your baited line out into the water. Easy.

There are many types of whip, but a 4-metre one is light and easy to handle.

A good whip has a length of *elastic* running through the inside of the rod, starting from the tip and going down the top section.

The elastic is brilliant at *absorbing* the fight from a fish.

Float

Split shot

Add a float, a hook and some split shot weights to balance the float. Together with your line, these make up your 'rig'.

Put your *bait* on the hook, then cast the whip out over the water and hold it still. . .

The rest is up to the fish.

RODS

A rod looks more complicated than a whip...

But when you combine it with the reel, you can cast out further and have more control bringing a fish to the bank.

The rod has a series of round *eyes* running along its length for threading through the line from the reel.

For beginner's float fishing, line with a **breaking strength** of around 5lb is perfect.

Float rod

Fixed spool reel

The *reel stem* is attached to the rod.

The *spool* is loaded with the main line.

The *handle* can be switched for left- or right-handed reeling.

The *drag wheel* acts like a clutch, helping you adjust the tension on the line and land a big fish that wants to run!

The *bail arm* is closed to trap line to the reel spool, and is open for releasing line to cast.

RIGS

The arrangement of all the bits you attach to the end of your line is your rig.

There are so many different rigs and techniques that it can feel quite overwhelming when you are first setting out. So, in this section I am going to focus on just one: a really simple *float fishing rig*. This was the rig I first learnt to catch fish on and I still use it all the time today.

Learning how to float fish well will teach you lots of useful skills that you can use throughout your fishing life, like casting accurately, understanding bites and fish behaviour, which bait to use, where to cast and why.

But first, let's start with the simplest rig you could ever tie: all you need is a *single hook*, and then you are ready to start freelining.

FREELINING

Freelining might sound fancy...

But it simply means fishing with nothing more than a hook tied to the line. It is as sensitive as fishing gets, and it really does work.

There are a few ways you can tell if you have a bite from a fish while freelining. You can place your fingertip on the line at the point it leaves the reel, or, if you don't want to risk accidentally moving the bait, just watch the line at the point it travels into the water. If the weather is calm, **look out for any trembles** or sharp movements, and strike when the line tightens or falls back (the fish could pick up your bait and run towards you!).

If the water is clear, you might be lucky enough to see the fish actually eat your bait, especially if you are freelining a *floating bait* on the water's surface, like a chunk of bread.

It is not unusual to hear your bait get eaten! **Carp** make this amazing lip-smacking 'cloop' sound when they take food off the surface; honestly, when you see a floating bait vanish into the mouth of a fish, it is rare for angling to get much more thrilling!

Freelining is a truly beautiful way to fish. It gives you a real sense of closeness to the fish.

For bigger fish you can use a larger hook and a bigger bait, or just add a few split shots to give the rig a bit more weight for casting further.

In flowing water, or if there is a wind blowing the water's surface in a favourable direction, you can even open your bail arm and gently feather out the line from your reel, rolling your freelined bait away from you and deep into a fish's lair.

Carp

YOUR TACKLE BOX

This is your box of tricks for fooling a cunning fish.

In your tackle box you can keep your lines, lures, baits, hooks – all the equipment required to catch a fish. You don't need a lot of tackle, but it is always good to have a few options for your day out – just in case the conditions or fish behaviour means that you need to make some changes to tempt a *bite*.

Having a few spares is always a useful thing too – no one wants to be going home because they've cast their one and only rig into a tree and lost it!

A strong waterproof box will also protect your tackle and keep it all in good order.

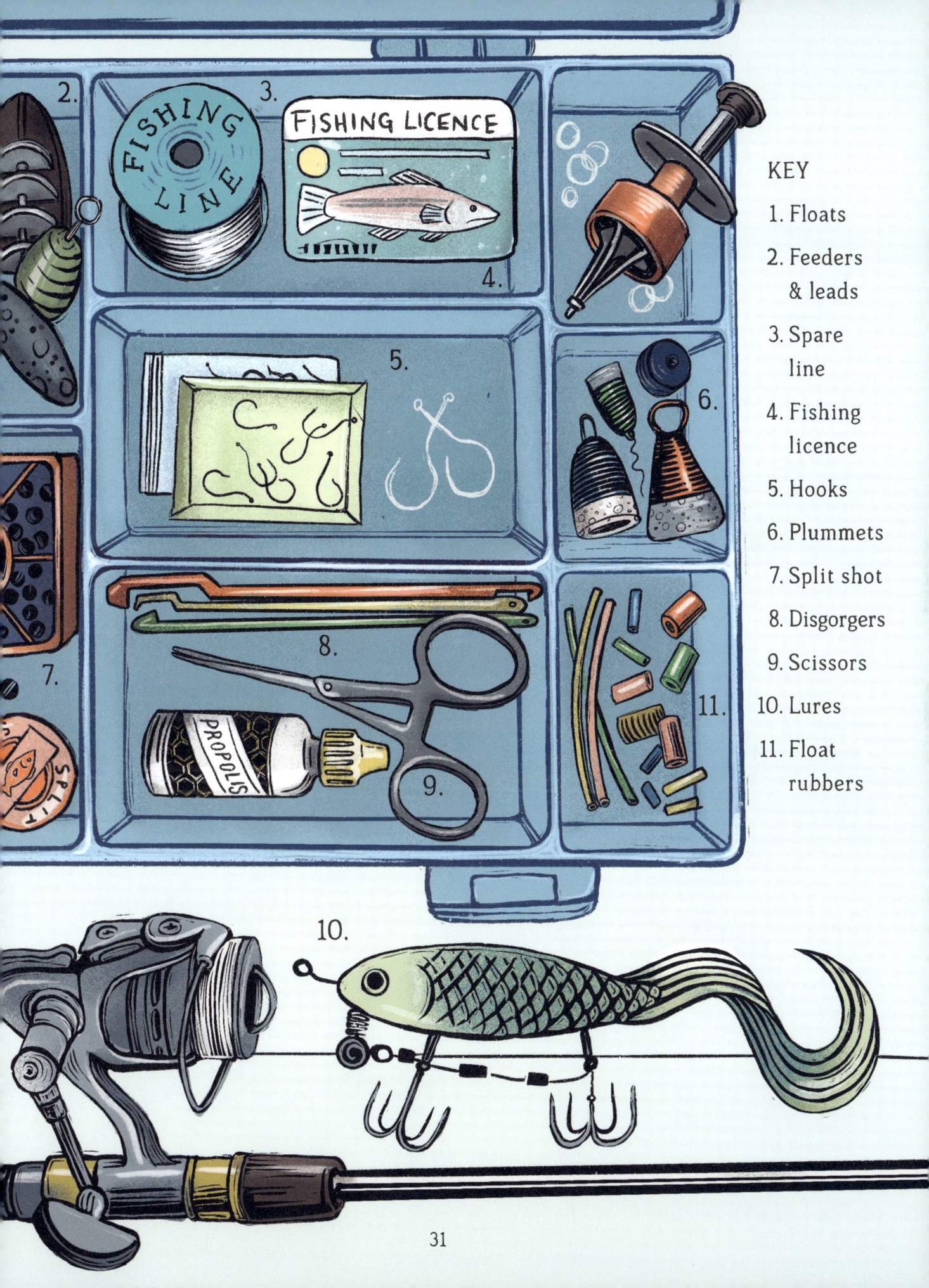

FLOAT RIGS

A float is a bright and buoyant device that tells you if you have a bite from a fish.

Floats suspend your bait in the water. They come in different shapes, sizes and colours – and they all do slightly different things. For this rig we are going to fish with what anglers call a *waggler float*. It can be used to catch a really wide variety of fish in lots of different fishing situations, and if I could only use one rig for the rest of my life, then this is the one I would choose.

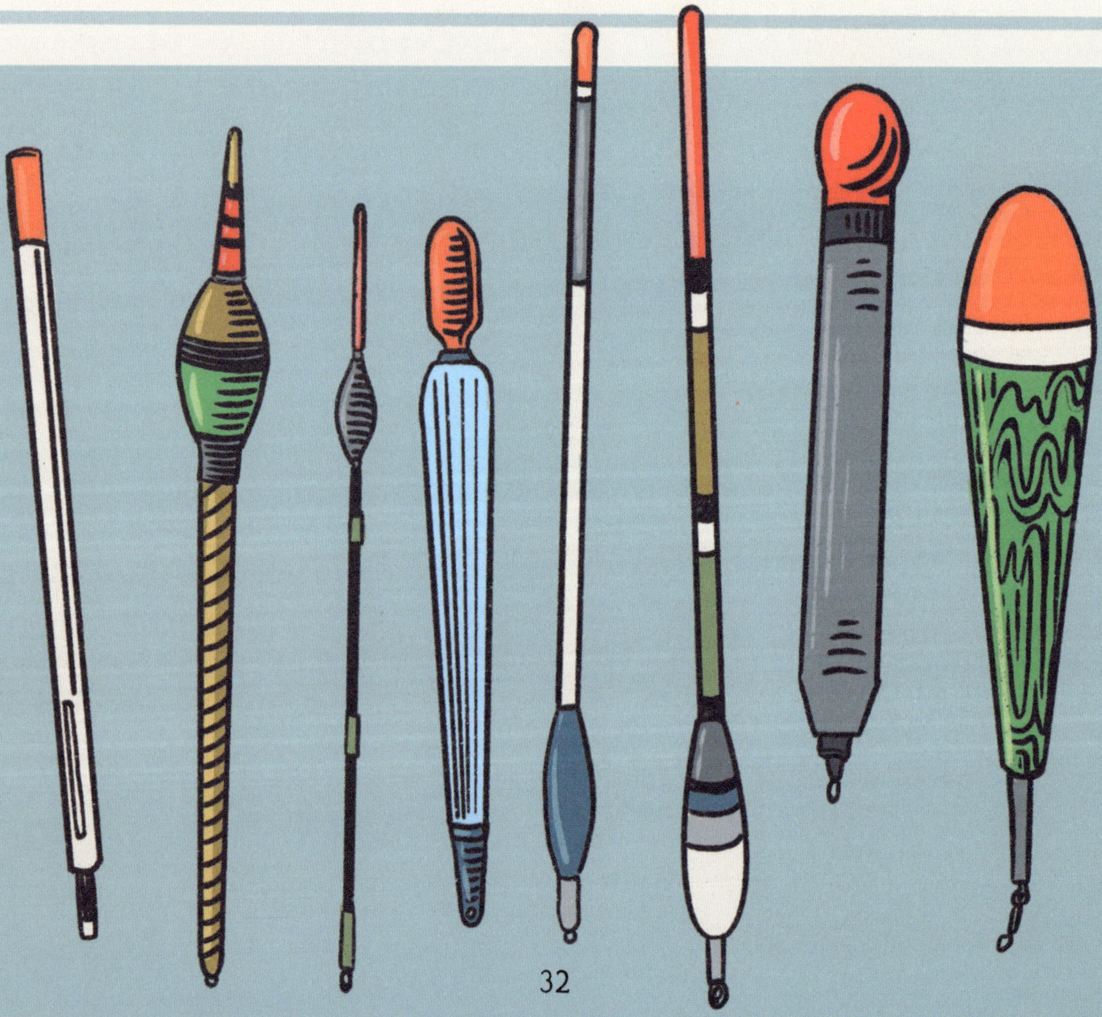

HOW TO ATTACH YOUR WAGGLER

1. The coloured tip of your waggler is the bit that sticks out of the water for your **bite indication**. At the other end, the float's 'stem', you'll notice a very small eye.

Holding your *waggler* in one hand, and the end of your *line* in the other, carefully slip the line through the eye as though you were threading a needle.

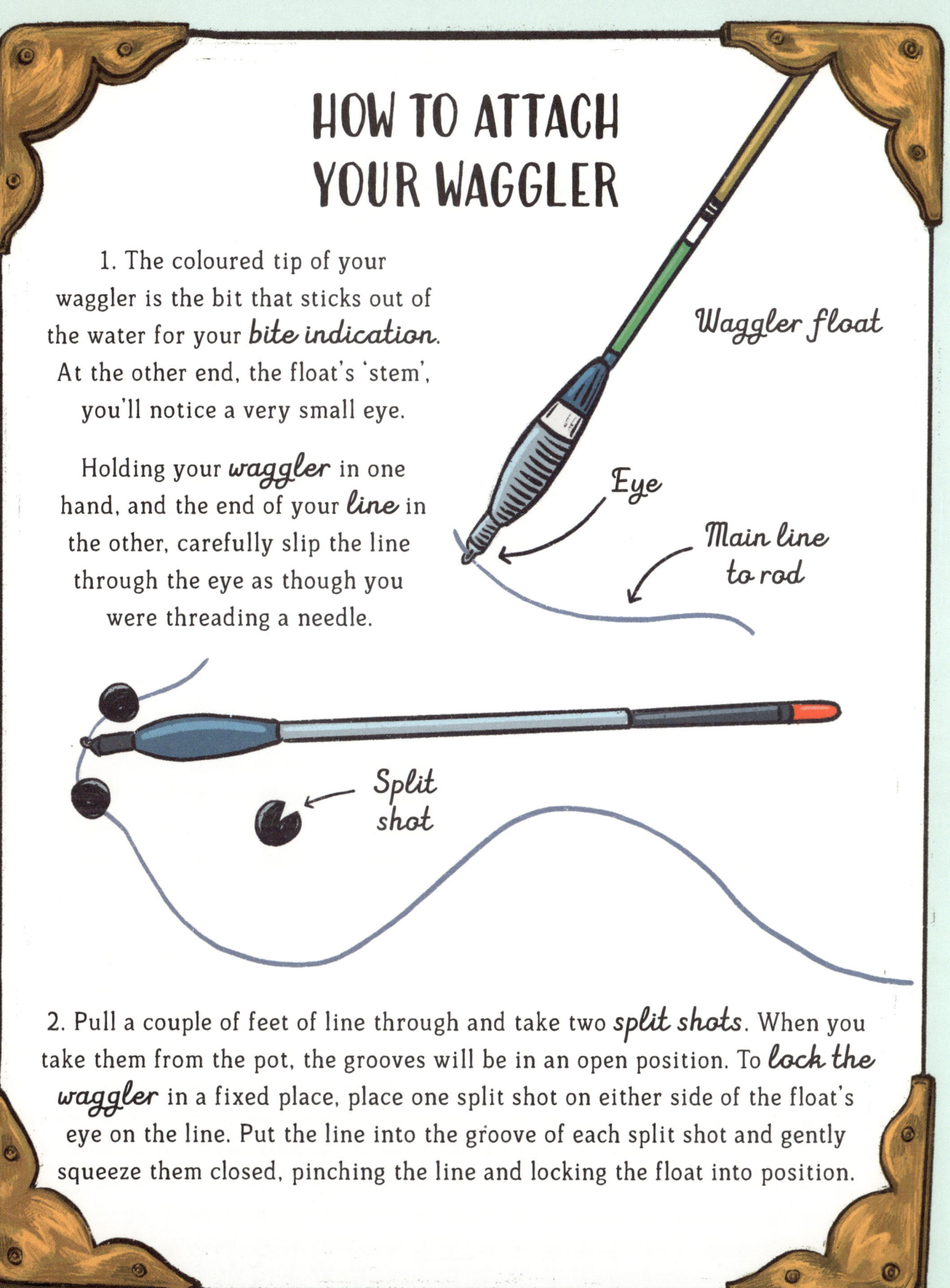

2. Pull a couple of feet of line through and take two **split shots**. When you take them from the pot, the grooves will be in an open position. To **lock the waggler** in a fixed place, place one split shot on either side of the float's eye on the line. Put the line into the groove of each split shot and gently squeeze them closed, pinching the line and locking the float into position.

HOOKS

Hooks range a lot in size...

At the top end there's size 1/0 for a monster fish, while size 22 is capable of catching a minnow. The one I suggest you pick is *size 16*, *eyed* and *barbless*: the perfect middle ground. The eye on the hook makes it more straightforward to tie, and we are going barbless because it is so much easier to get it out of a fish's mouth (or yourself, should you be unlucky enough to hook your own finger!).

HOOK SAFETY

Casting with rod and reel means you'll have that hook flying through the air. You can easily hook people behind you – I've had many hooks go into the back of my ears and head. The risk of that happening can be reduced to almost zero with the *good casting techniques* we will learn later in the book.

Sometimes, if you're unlucky, you might hook a tree on the far bank, or get your hook *snagged* on the bottom of the river or lake bed. Most of the time, just a bit of *pressure*, or a change in *angle*, will shift it; or at least snap the line close to the hook.

A CAUTIONARY TALE

When I was eight, I hooked the bottom of my local river. I had huge barbed hooks on my line. I pulled hard, they popped out of the snag and hit me directly in the eye, sending me straight to hospital. By pure luck my eye recovered, but I have got a permanent scar across that eyeball. Many have not been as fortunate, so *please be careful*.

KNOTS

Never forget: your knot is the weakest part of all your tackle.

The best bit of advice I can give you is to practise. You can buy pre-tied hooks called *hooks to nylon*, but ultimately you will need to tie a knot at some point as you progress in angling. It is much better to get it all nailed down right from the start. Here are three classic forget-me-knots.

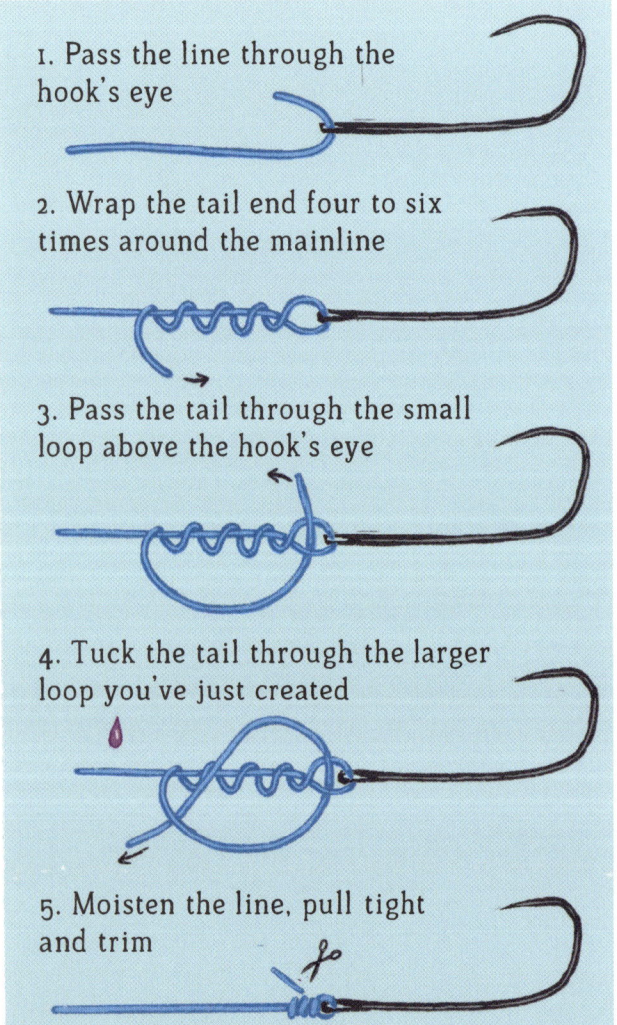

1. Pass the line through the hook's eye
2. Wrap the tail end four to six times around the mainline
3. Pass the tail through the small loop above the hook's eye
4. Tuck the tail through the larger loop you've just created
5. Moisten the line, pull tight and trim

Half-Blood Knot

The half-blood knot is a classic and will set any angler in good stead.

Before you pull any knot tight, always wet it with spit. This reduces the friction (and the chances of a break) and helps the knot to slide together nicely too.

If your knot does not look perfect, then cut it off and start again from scratch. A bit of lost time hurts far less than a lost fish.

1. Make a 6-inch loop in the tail of your mainline

2. Pinch the loop's neck with one hand, then create a second loop by tucking your first back behind your standing lines

3. Bring your first loop back through the second, forming a loose figure of eight

4. Moisten and pull tight

5. Attach a hook length loop-to-loop by passing your hook length loop through the mainline loop, then your hook through the hook length loop, and pull it tight

Albright Knot

This is one of the best knots for connecting two sets of line where one is thicker than the other.

1. Form a loop in the thicker line and pass the thinner line through the loop

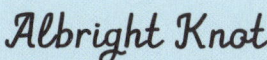

2. Neatly wrap it six times around the loop

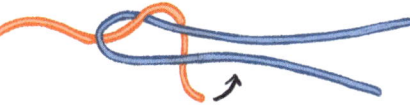

3. Pass the tail end back behind the loop's tip and through the hole

4. Moisten the line and pull tight

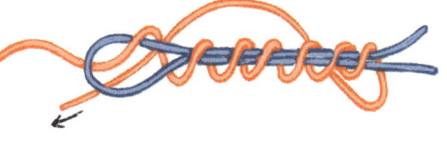

5. Trim

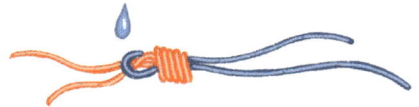

Figure-of-Eight Knot

Learning the way to tie a strong loop in your line which won't slip is really useful for attaching hook-lengths or for attaching a whip connector.

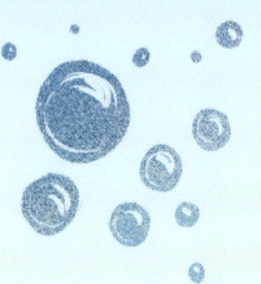

PLUMMETS

Knowing the depth of the water is really important.

It allows you to start forming a picture of what that mysterious underwater world actually looks like... and where the fish might be *feeding*.

So, with your waggler locked in position, and your line bare down to your hook, we are now going to use a little device called a *plummet* to measure the depth of the water.

Once you have that information, it's a great idea to then move your float rig and plummet around a few other areas in your fishing spot to see if you can find *underwater features*; there might be a step into deeper water (a great fish-holding area) or a gentle slope, a sudden deep hole or a surprisingly shallow bar. All of these are great things to notice and make a note of.

We are going to look at places where you might find feeding fish a little later on, but just remember that fish can move and feed at different depths, so knowing the *maximum depth* is a huge advantage.

HOW TO USE YOUR PLUMMET

Slip the hook through the *eye* at the top of the plummet and then hook it directly into the *cork rim* below. Cast it out into the area you want to fish. The split shots locking the waggler can either be pushed up and down the line (if you haven't put them on too tightly) or opened and moved, to make all the little adjustments to your depth.

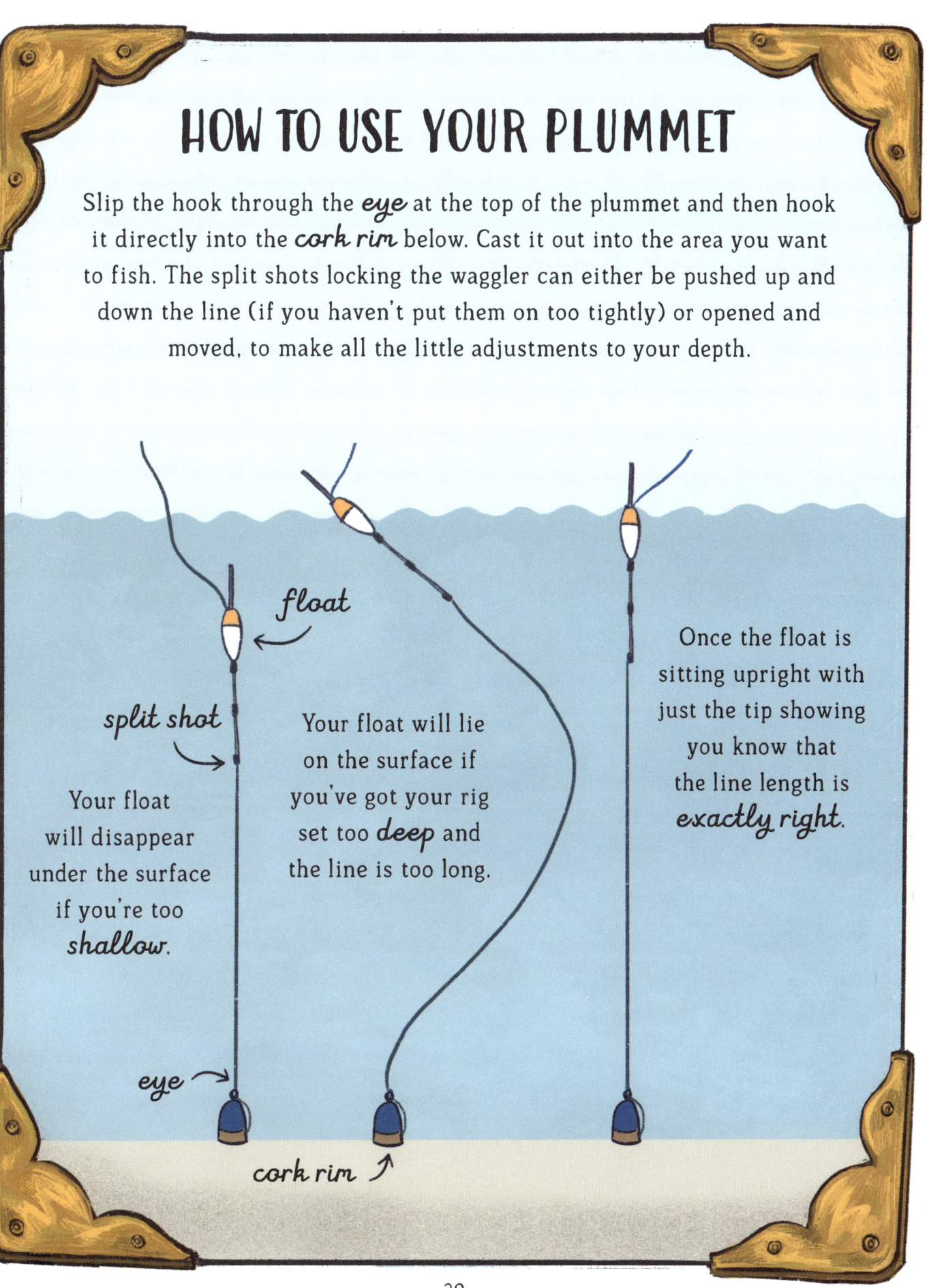

float

split shot

Your float will disappear under the surface if you're too *shallow*.

Your float will lie on the surface if you've got your rig set too *deep* and the line is too long.

Once the float is sitting upright with just the tip showing you know that the line length is *exactly right*.

eye

cork rim

SHOTTING UP

The final pieces of tackle to complete your rig are the split shot weights.

There are a few reasons why they are important. Firstly, you want your float to be as *sensitive to bites* as possible. If it hasn't got any weight to keep it upright in the water, it'll just lay flat on the surface and not tell you much; too much weight, though, and it'll sink right out of sight.

Having weight on the line helps you to cast your rig out with accuracy, and avoids tangles too. Perhaps most importantly of all, though: *different shotting patterns* will help you to find the feeding fish at the different depths of the water you are fishing.

SHOT DISPENSER

This wheel is filled with different sizes of split shot. The sizes are labelled with numbers or letters.

The biggest shot might be an SSG, then AAA, then BB.

After that, the number system takes you down to 12 – a shot so small that I struggle to hold it, let alone pin it onto my line using the tiny little groove!

CALCULATING YOUR SHOT

On the side of your waggler float, you'll see it tells you exactly how many shot you need to balance the float perfectly.

For example, if it says 3 x BB, it means you need three of the shot labelled BB to get that float sitting correctly in the water.

However, that doesn't mean you just stick three BB on and cast out. You will put the bulk of the shot around the float stem, to help with the cast, but you'll then need smaller shot spread down the line, below the float, to help send the baited hook down through the water.

Here's a really useful chart to help you figure this all out.

SIZE	WEIGHT	EQUIVALENT
SSG	1.6g	2 X AAA
AAA	0.8g	2 X BB
BB	0.4g	2 X No. 4
No. 4	0.2g	2 X No. 6
No. 6	0.1g	2 X No. 10
No. 8	0.06g	3 X No. 12
No. 10	0.04g	2 X No. 12
No. 12	0.02g	2 X No. 13

SHIRT-BUTTON PATTERN

A good place to start with a 3 x BB waggler is to attach two BB around the stem of the float to lock it into position. To make up the weight of that final BB, you could then evenly spread four more *progressively smaller split shots* on the line below, in a shirt-button style. One option could be: one No. 4, followed by one No. 6, then one No. 8, and finally a single No. 10, pinched just a couple of inches from the hook.

With this set-up you are far less likely to get tangles on the cast, as the float – the heaviest part of the rig – will hit the water first, followed by the rest of the line and then the hook. The shirt-button grouping of the split shots will also allow your bait to *flutter through the water*, instead of just sending it straight down, and the final, lightest shot will help set the hook into the mouth of the fish.

DEEP DRAIN FISHING

As I got older, and a bit better at casting, my grandad started taking me to a deep drain to fish. There, the depth could be 10 feet, and I soon learnt the bigger fish grouped around the final foot or on the absolute bottom itself.

I would set my float at 10 feet deep and then put the bulk of my shot about a foot from the hook. That *shotting pattern* meant my bait would fall quickly through the water, without getting picked up by smaller fish on the way down. It would then hold my bait in the perfect spot without it being moved around too much.

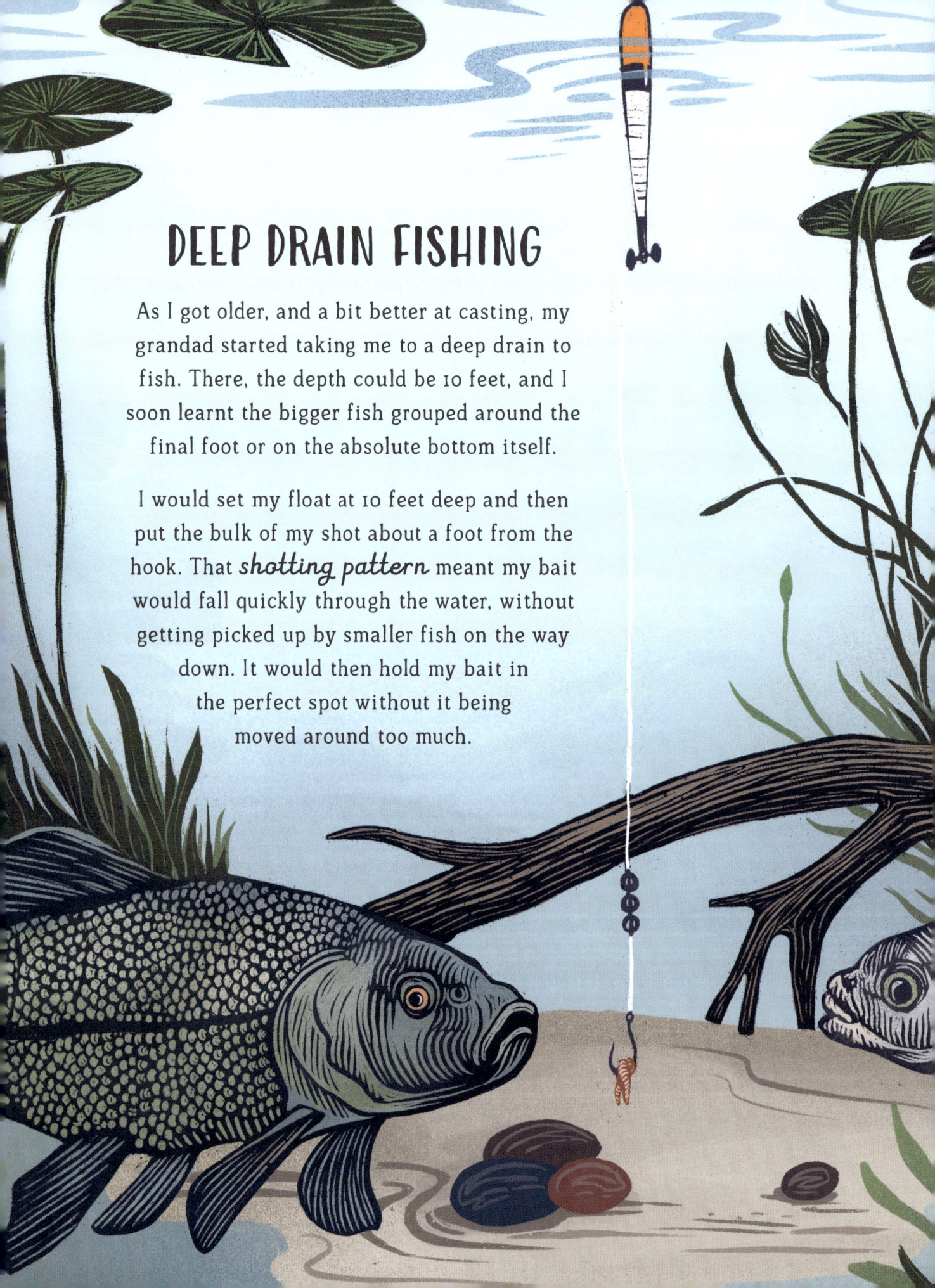

SHALLOW FISHING

Around the same time, I also got good at fishing in really shallow water; often just 6 inches deep. A lot of the fish I caught would be *on sight,* meaning I would see them, cast over them, and then slowly reel my waggler and bait right back in front of them.

I wouldn't want my bait to sink quickly, so I would bulk the shots around the waggler stem and leave just one tiny split shot near the hook, or, I would use no shots below the float at all. For catching *surface-feeding* roach and rudd, it was absolutely deadly!

OTHER TACTICS

Once you feel confident with your float rod, why not try a few different methods?

The wider you grow your base of skills, the more places you'll find yourself able to fish, and the greater number of fish species you'll catch. You will need different rods, reels and line strengths for the methods below, and some extra bits of specialist tackle (like a pike unhooking kit for predator angling), but your local tackle shop will be able to help you.

Legering and feeder fishing involve using a *weight* to anchor your *bait* to the bottom of the river bed or lake, and can help you to catch bigger fish like carp, bream and tench.

Barbel

LEGERING AND FEEDER FISHING

Depending on how heavy the weight or target fish is, you will probably need a *stronger*, and stiffer rod and *thicker* line. Legering uses a single weight called a *lead* or a *bomb*, which is attached directly to your mainline or rig. They come in different shapes and sizes depending on how far you want to cast or the fishing situation. For example, in fast-flowing rivers you might use a gripper lead, which is flattened down and grooved to help it *grip the river bed*; or, for long-distance carp fishing, you could use a heavy lead that's more rocket-shaped and aerodynamic, to help it *fly further* through the air.

Feeders take the principle of the legered weight and allow the angler to add attractive free food to bring the fish into their fishing area – whether that's maggots crawling out of the holes of a maggot feeder, or sticky sweet-smelling breadcrumbs in your groundbait feeder!

bait

weight

PREDATOR FISHING

Something monstrous... Something deadly... The water wolf.

I can remember when I first became aware that something else was lurking in the little river outside my home.

I was float fishing to catch small silver fish. The float dipped, but as I reeled in my fish, an enormous pair of *jaws* erupted from the gloom behind.

As I leapt, the little roach I'd hooked was absolutely engulfed: hook, bones, fins and all. My line snapped and all I was left with was a battered float rig hanging limply from my rod tip, and a trail of tiny silver roach scales disappearing into darkness. It was the *water wolf*. The pike.

And from that moment on, I was obsessed.

FLY FISHING

Quite honestly, I have to tell you that I am a very bad fly angler...

But I do love fly fishing because I believe it teaches you about fish, the water, and especially the insects that fish eat. Watching an expert fly angler at work on a wild stream is like watching a fine *artist* as they paint their masterpiece.

So how is it different? In regular fishing, we weight our line with the rig we tie and the bait we use. In fly-fishing, there is no weighted rig, split shot or lead, and the fly bait is super-lightweight – the weight to cast out comes from the special type of weighted mainline that fly anglers use on their reels.

The *fly* itself is made to imitate those very insects: sometimes larvae or worms, but mostly a form of flying insect (one that might have recently hatched on the water an angler is fishing in).

BAITING UP
MAGGOTS

Once you've discovered how brilliant live maggots are as bait (trust me!), you will learn to *absolutely love them*.

Maggots are my number-one bait for anyone new to fishing, and definitely the bait I use whenever I am fishing somewhere for the first time. They will catch virtually everything that swims. Fish with one or two on a small hook for little silver fish, or use a bunch of half a dozen or more on a big hook to catch larger fish.

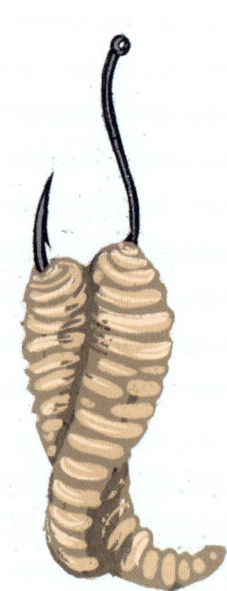

You can buy *maggots* by the *pint* at your local tackle shop.

Nick your hook through the *blunt* end of the maggot.

Throw six to twelve maggots in the water as a *loose feed* every five minutes to get the fish confidently feeding on your spot.

You can store maggots to use again another day: they keep for about a week. The *fridge* is best, but please ask permission first and make sure the lid is on properly! I once put maggots in our fridge without asking, only for them all to *escape*. . . It was one of the worst days of my life.

WORMS

The *humble worm* has probably caught more fish worldwide than any other bait in history. Not only are they free, worms also wriggle enticingly on the hook and let out juices that drive predatory fish like perch and chub crazy. All kinds can be used: from the night-crawling six-inch lob worm to the tiny red or banded brandling worm.

Find them under old *logs*, dig around the garden or compost heap, or scour the grass right after a rainstorm.

Store worms in a *tub* of damp earth topped with moss or wet paper shavings. Punch breathing holes in the lid and feed them vegetable peelings. Keep them cool, change their bedding and they will live for many months.

BREAD

A basic loaf of white bread is a cheap but brilliant bait. To get fish feeding near or on the water's surface, tear off large pieces of bread and throw them straight in. If you want to entice the fish to feed nearer to the bottom, lightly *dampen* a slice and *squeeze* it tight to make it sink.

Fold a chunk around your hook (leave the point clean and proud), then pinch at the top of the hook's stem to *pin* it in place.

Bread isn't just for small fish – *huge* carp and chub have been caught with a large chunk!

SWEETCORN

Sweetcorn: what a *brilliant bait*. Sweet-smelling, bright, durable and easy to buy. For many years, a humble kernel of corn was responsible for my biggest freshwater fish, but I was far from the first person to realize the superpowers of sweetcorn.

On 16 June 1980, legendary angler *Chris Yates* broke the British carp record when he landed an enormous 51.5-lb fish called *The Bishop*. Despite all the advances in modern bait and tackle, he did it with nothing more than a hook baited with three grains of corn and a lump of plasticine as a weight.

TINNED MEAT

This is another *inexpensive bait* that you can buy anywhere. Luncheon meat, spam, tinned hotdogs and pepperoni sausages can all be cut into pieces that fit on your hook and last longer than softer baits like bread. The meaty smell attracts fish of all sizes.

BOILIES

These small, round balls are filled with attractants and come in all sizes and colours.

A boilie is set to your hook using a device called a *hair rig*, which is a little complicated for first-time anglers, but the major benefit is that it lasts a a long time in the water and is difficult for small fish to eat, which helps anglers hoping to catch larger fish like carp.

GROUNDBAIT

Groundbait is usually a type of finely *ground breadcrumb* with colours and flavours that, when mixed with water in a bucket, can be rolled into firm balls and thrown into the water to form a delicious cloud of treats to bring the fish in.

ESSENTIAL EXTRAS

Wear *good outdoor shoes* and *check the forecast* before you leave. Bring a raincoat and wellies if it is going to be wet; a warm jumper for the cold; sun cream and a hat if it is hot; and a head torch if you are going to be out before sunrise or after sunset.

Sun cream: Water reflects sunlight into your face, so even with a hat on you can get burnt!

Mobile phone: Useful for emergencies, photographing your catch and keeping an eye on the time!

Fishing umbrella: Great for both very hot and really wet days.

Basic first-aid kit
Hooks are sharp! Bring plasters, bandages, medical tape, tweezers and antiseptic.

Bait box: A plastic tub with small holes punched in the top so your living baits can breathe. Use a bait box to store all your bait for the day.

Unhooking mat: A padded cushion to lay your fish on while you remove the hook. Make sure it's damp. Fish can lose their protective slimy coating if handled with dry hands or left on a dry mat.

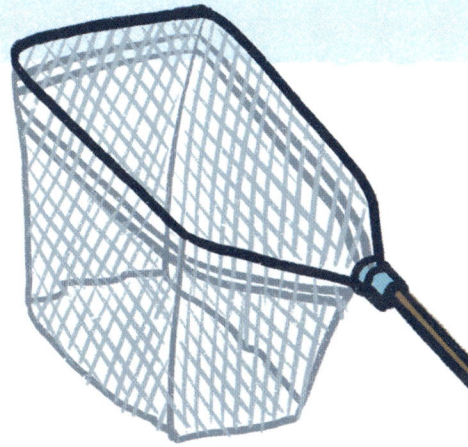

Landing net: For scooping the fish on the end of your line from the water safely on to the bank.

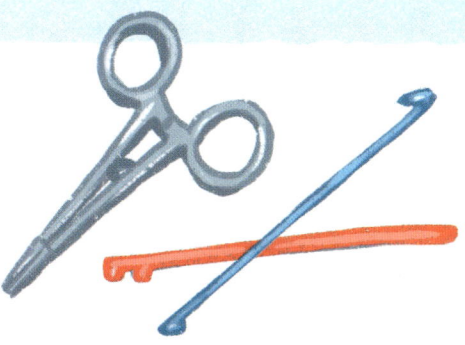

Disgorger or forceps: Tools for quickly and safely removing your hook from the fish's mouth.

Food and **drink**: Anglers who are well fed and well watered definitely fish better!

KEEPING OUT OF TROUBLE

It's important to fish legally

First, check if you require a licence to fish. Where I fish – in Wales and England – if you are under thirteen you do not need one, and if you are aged between thirteen and sixteen you will receive one for free on application – just head to the official gov.uk website for all the details.

Next, get permission from the bankside landowner before you fish. Fishing clubs will have sorted all that for you already; so, for the cost of a club book, you will have access to a whole range of waters. Plenty of fishing spots sell day tickets too, especially commercial stocked fishing lakes.

If you are feeling a bit adventurous, get a map out and investigate your waterways on foot first. Some might be free to fish (check online), or a bit of door-knocking and polite persuasion might do the trick.

It's important to fish safely

It is essential that you stay safe. Tell people where you are going and let them know when you plan to come home. It can be really hard to drag yourself away, especially if the fish are biting, but if you have made a plan then stick to it.

Being able to swim confidently is crucial. Always respect the water in front of you: it can be surprisingly deep in the edges and often very cold, even when the sun is shining. You never know what obstructions might be under the water by your feet (sometimes, whole trees are submerged down there), and underwater currents, even in knee-deep water, can be more than enough to sweep you off your feet, out into deeper water and even deeper trouble.

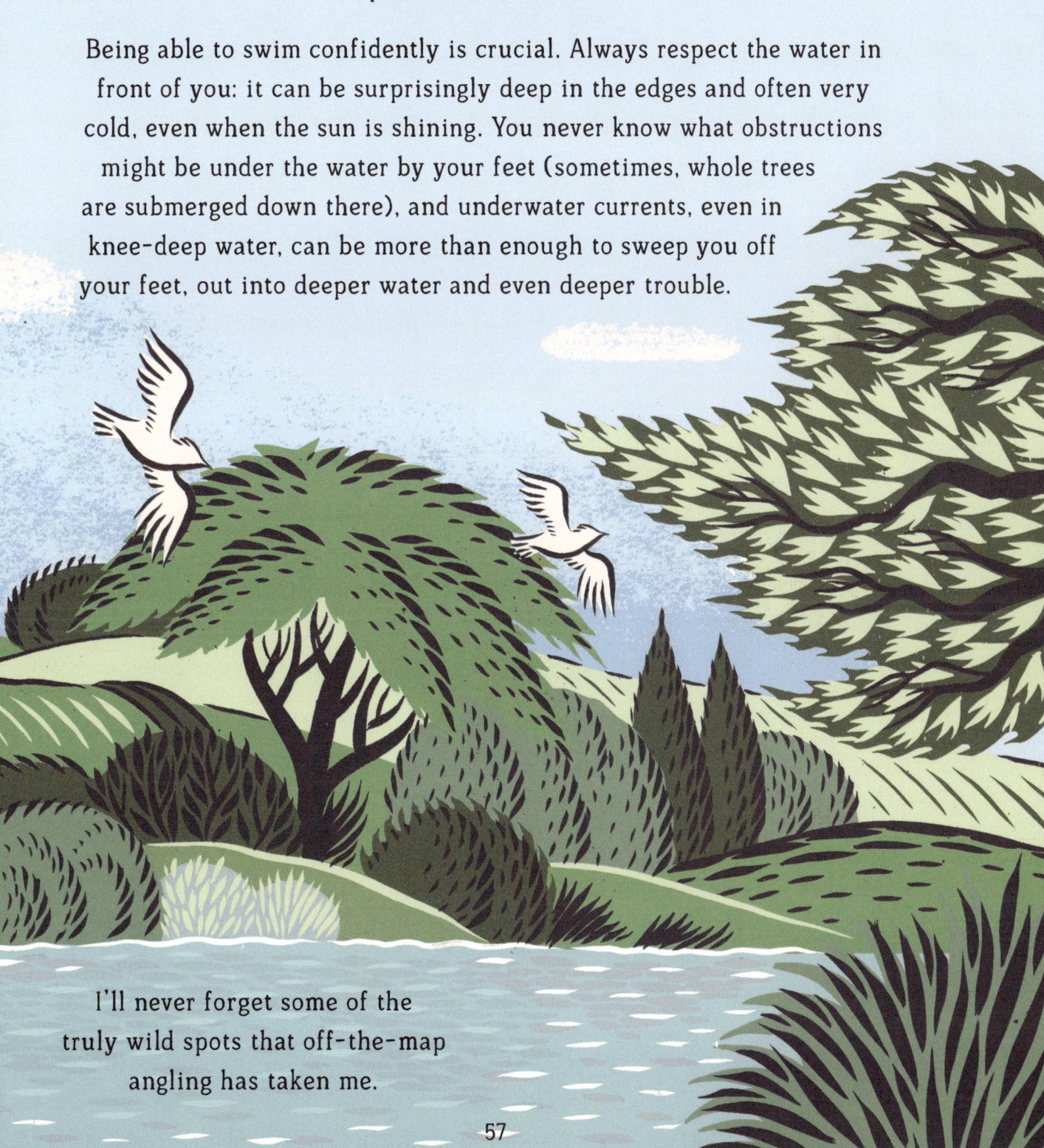

I'll never forget some of the truly wild spots that off-the-map angling has taken me.

THINK LIKE A FISH

Okay, you have everything you need. Your rod is ready.

What next? Let's take this right back to the most obvious thing: if you want to catch some fish, you need to go where there is water, and once you're there, you need to think like a fish. And what is most important to fish is that the water they live in is clean. Clean water doesn't always mean clear water, but it must be unpolluted. Check the water isn't smelly, oily or greasy-looking.

Scan the banks for nests and small holes made by creatures like water voles.

Train your eye to look for the right things and you might be really surprised at the sorts of places where really great fishing spots can be found.

Fish don't care what their home looks like. I have seen fish in their tens of thousands in forgotten city-centre canals. In fact, the biggest freshwater fish I have ever seen – an absolutely giant five-foot-long catfish – actually lived inside a pipe on the edge of a very busy town.

Some good clues to look out for when searching for a place to fish include a decent mix of plant life: weeds, rushes or lilies.

Can you spot the footprints of visitors like otters, stoats, weasels or badgers? Are there waterfowl such as ducks or geese, or fish-eating birds like herons, grebes and kingfishers? All of these are excellent signs that fish lurk below.

The water's edge often acts as a nursery for young fish.

Try to spot tiny critters like pond skaters, water boatman or dragonflies.

WATERCRAFT

One cast in the right place is worth more than a whole day casting in the wrong place.

I know how exciting it is when you fish somewhere new, but slow down and take your time. Fish have a sensory system along their sides called the lateral line, which feels any noise as a vibration and helps them to *flee from danger*. So, as you approach the bankside, imagine you're a cat creeping up silently on your prey. One misplaced step, dropped bucket or raised voice could scare off the fish for ages.

The best thing you can do is to discover where *fish are feeding*. See the fish without them seeing you. Make sure you have some cover behind you, like foliage or a canal-side wall, or simply stay low to the ground as you quietly peer in.

In clear water, you might see the fish straight away. Study them and look for fish that seem active but relaxed, or, ideally, fish that are *clearly feeding*. Some fish have sites where they will feed at over and over again. You can lay your baited hook out as a trap, ready and waiting for their return.

Sometimes you can't see a fish, but you know it is there from other clues. *Rings* or *swirling patterns* on the water's surface are signs that fish might be feeding there. You might hear splashes or a '*cloop*' sound made by big fish as they ambush insects.

If part of the water looks *cloudier*, or pockets of *bubbles* appear, it might have been stirred up by bottom-dwelling fish-eating bloodworms. Study the movements of the plants at the water's edge too. *Twitches* and *trembles* could be caused by a fish down at a plant's base, hunting for freshwater invertebrates.

Look for *water birds* and where they feed: ducks and swans sucking up blanket weeds can stir up the insects, so fish soon learn to follow waterfowl for an easy meal. Fish-eaters like kingfishers, herons and crested grebes will scare the fish in the areas they hunt – and if small fish leap out and spray across the water in panic, you know that somewhere down there is a big *hungry predator* giving chase too.

Perhaps the best creature to seek out is another *angler* who knows the area. You can learn so much from a polite enquiry – and if you see them catching, they are probably going to be in a good mood, too. We all had our first day at some point, and all decent anglers will help each other out, especially beginners.

CASTING OUT

Finally. You are on the bank, ready to start.

Think about how you are going to land your fish safely. Avoid a cluttered or high-up bank, where it can be *dangerous* to lean out across the water to scoop a fish into your net. No fish is worth risking your own safety for.

Set up your landing net, mat and unhooking gear first. Now it's time to *cast*! Casting out is probably the trickiest part of your first day of fishing. Tangles and lost rigs in trees and bushes can be frustrating. Remember, you don't need to cast far to fish effectively. Fish can feed inches from your own bank! In those cases, you hardly need to cast at all – you just *ship your rig out* and carefully plop it in.

Both these casts require you to first trap the line on your reel behind the fingertip of your index finger – and then open the bail arm like this:

Trap line behind finger tip first, then open the bail arm

Grip your fingers either side of the reel stem

UNDERARM FLICK

For *close casting*; useful when there are trees and bushes behind you!

1. Lift your rod tip up and swing your rig towards you

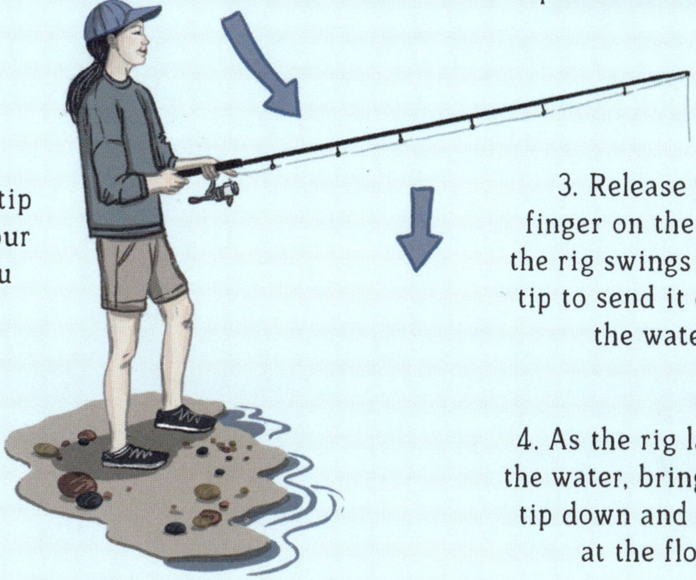

2. Gently swing the rod forward, bringing your rig past you in a pendulum motion.

3. Release your finger on the line as the rig swings past the tip to send it out into the water.

4. As the rig lands on the water, bring the rod tip down and point it at the float.

OVERARM CAST

For *casting further*: practise! No one gets it right first time. You can try it on your lawn at home.

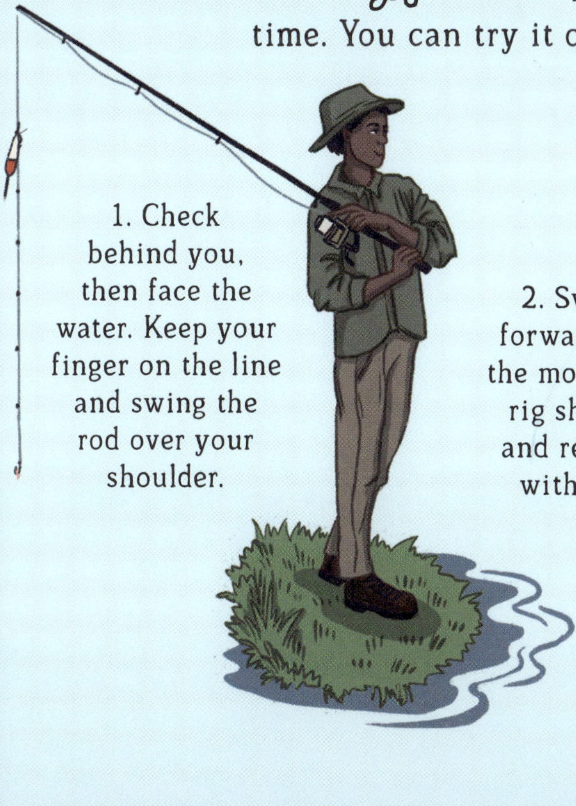

1. Check behind you, then face the water. Keep your finger on the line and swing the rod over your shoulder.

2. Sweep the rod forward, then break the movement as your rig shoots forward and release the line with your finger.

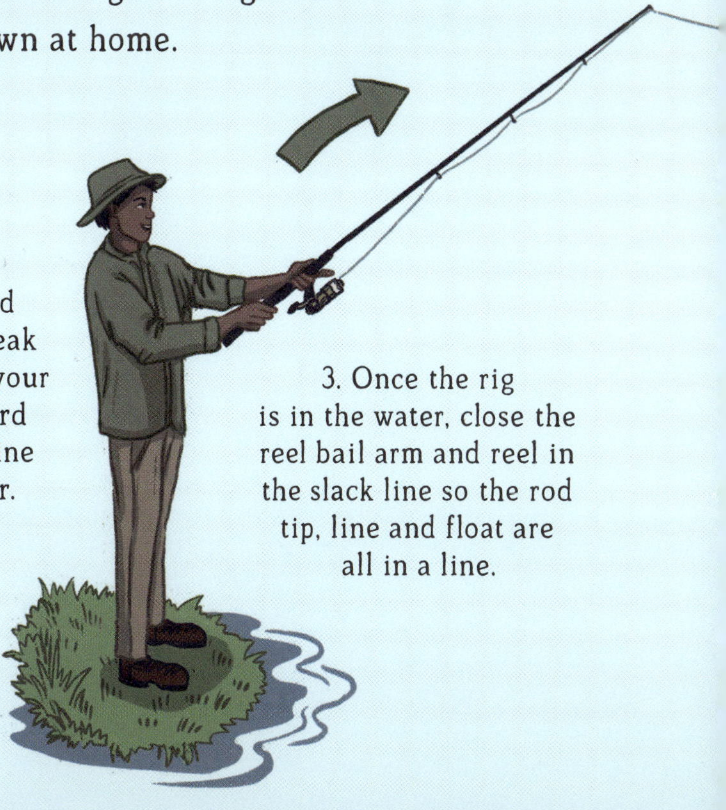

3. Once the rig is in the water, close the reel bail arm and reel in the slack line so the rod tip, line and float are all in a line.

WAITING

Anticipation is everything in angling.

All that building tension before the bob on your float and the action begins can be really exciting – but you might not feel anything right away. That doesn't mean you won't catch a fish. Sometimes you just need to make a small change: a different *bait* or, if you are float fishing, just *experimenting* with the depth you are fishing at. Introduce a little groundbait or loose feed. Still nothing? Don't worry. Angling is about *exploring*. If you can, move on to the next spot and start again.

Be confident and stay *positive*. Occasionally we spoil our chances by making too many changes to our tackle, disturbing the water by recasting too often, or putting out too much groundbait or loose feed.

Most fish feed in spells, and very often their preference is dawn or dusk – when the *gloomy light* makes them feel more confident coming out of hiding. The weather can also play a huge role. A little cloud cover over the sun can create the right sort of shade when it is mild, and in a really cold period, a bit of *sunshine* can provoke them into biting.

One golden rule, though: never ever fish if there is any *thunder* and *lightning* around. Your rod is an excellent conductor of electricity and anglers have lost their lives in the past due to lightning strikes on their rod. Don't take the risk.

Knowing when to go *home* is important. If it just isn't your day no matter what you do, even a day without a fish is always a day to learn something new, and it is always a lot better to be out fishing than being stuck indoors. Some of my most memorable moments on the bank have been when I caught nothing. Sometimes it can be about netting an amazing fish for a friend; other times you might witness some special wildlife sights on the water.

WIND

When checking the forecast, look at the direction of the wind.
There is an old *angler's rhyme* that goes:

When the wind is in the west, the fish bite the best.
When the wind is in the south, it blows the bait into the fish's mouth.
When the wind is in the east, the fishes bite the least.
When the wind is in the north, the fish don't go forth.

Truthfully, I've caught in all winds, but there is something about a harsh easterly that they definitely don't seem to like!

STRIKING A BITE
The float is perfectly balanced...

You've been staring at it, willing it to move for so long, that when it suddenly does you almost can't believe it. The first time my float *disappeared* under the water, I was so surprised and struck so hard that I sent my first rod-caught fish flying out of the water, over my head, deep into a field filled with sugar beet.

The key is to stay *calm* and time it just right. You very rarely need to strike hard, especially if you are fishing in close with a float and small hook. As long as the line between your rod tip and float isn't slack, just a sharp, but quick, flick of your wrist is more than enough.

It is so easy to strike too *early*, at the very first indication of a fish, or strike too *late*, after the fish has already ejected the hook from its mouth. Timing your strike is not an exact science and does take practice, but it is definitely better to be too early than too late, and risk the fish swallowing the hook down deep.

1. The float bobs under as the fish eats the bait and swims straight down

2. The float lifts up as the fish eats the bait and swims straight up

3. The float glides sideways as the fish eats the bait and swims off without going up or down!

Kneel down low, submerge your net, guide the fish back over the net, and lift!

LANDING A FISH

The fish is on! If you are fishing for small fish, you should be able to reel and lift them quite easily. But if the fish is too big to lift straight out, or too powerful to reel straight in, then you will need to *play* the fish – one of the most exciting and nerve-wracking parts of fishing! Use your rod to cushion the fish's fight and adjust your reel's drag to allow it to take line if it needs to run!

FISH SAFETY

Your first fish is in the net! The next job is safely lifting and carrying your fish.

Unhook a small fish in the net by the water's edge. If you need to handle it, always make sure you are holding it above the net or over the water in case you *drop* it.

Use your net to carry a bigger fish onto your landing mat, checking its *fins* and *gills* are flat against its body – not bent awkwardly or trapped in the net's mesh – and its *jaw* (if it's hard) isn't tangled in the net's mesh.

Make sure your mat is *wet*, and always handle your fish with wet hands. Dry surfaces will remove a fish's protective *slime* coating or its scales.

Keeping your fish cool ensures it remains calm – a small bucket of fresh water is really useful to have close at hand. The key is to handle the fish as little as possible and get it back to the *water* quickly.

REMOVING THE HOOK

If the hook is in the corner of the mouth or the lip, remove it with your fingers. Never pull it. Instead, carefully ease the hook out by pushing it back the way it went in. *Barbless* hooks make this much easier to do.

If the hook is deeper, you'll need to use your *forceps* for a large fish, or a disgorger for a little one. With the forceps, look at the position of the hook, then guide the forceps inside the mouth, grip the hook and remove it as above – as if the tool was an extension of your fingers.

To use the *disgorger*, first put a little bit of tension on the line above the fish's mouth, then place the line into the disgorger's slot and twist it once to hold it in place. Gently slide the disgorger down the line, into the fish's mouth and down onto the bend of the hook. Apply a little downward pressure and the hook will *pop out* – then just carefully remove the disgorger, with the hook attached, clean out of the mouth.

Rarely, but sadly, the hook is just too deep to remove. If that happens, cut the line as close to the hook as you can. Never attempt to pull a deep hook out of the mouth – you will definitely cause far more *damage* to the fish and could easily kill it. Hopefully, the hook will fall out on its own – if you've used a barbless hook, the chances of this happening are much higher.

CAPTURING THE MOMENT

If you want to *weigh* your fish and get a photo, always think about how long it has been out of water. If it is a really hot day, it is better to just take a picture of the fish in the water in the net, and put it back without the extra stress of a longer trip in the air.

If the conditions are right to take a picture, keep an eye on the time: this is a job that should be done in around a *minute*. Once the hook is out, you can always rest the fish in the water's edge in your *landing net* – giving it some time to recover while you set up for a photo or prepare a weigh sling. To take a picture of your catch, ask a friend to manage the camera, or get a tripod with a 'self-take' feature or a timer, so you can concentrate completely on handling the fish safely. Wet your hands again and *cradle* the fish beneath its body without gripping it too hard and damaging its organs.

Remember: out of the water there is nothing to support the fish's weight, so make sure you've got both hands evenly spread under the fish, and never just grip the lip, head or tail on its own. Always be ready for the fish to *flap* or slip from your hands. Keep your landing mat beneath the fish and kneel down for the photo: don't ever stand, as a fall from height can badly hurt the fish. With a bit of experience, you can often feel when they are going to flap – when they do, just gently lower them back to the mat to relax and try again.

SAYING GOODBYE

In the UK, there are only a few places where you can kill a freshwater fish to take home to eat – and they are mostly stocked trout ponds. The sad truth is that the *fish populations* here just couldn't cope if everyone took their fish home to eat.

Make sure the fish is definitely ready before letting it go. Gently lower it into the water. Most of the time, your fish will probably swim off with no problems at all, but if it is tired then you will need to help it *recover*. If you caught your fish in flowing water, support it in the net and ensure it is upright, with its head facing upstream. This will let the water run through the gills of the fish, giving it the *oxygen* it needs. In slow water and ponds, just rest it in the water's edge using the net. Once the fish is ready, you'll notice it starting to kick a little with its tail and nose at the net – then just ease it out to swim back to its home.

DIFFERENT FISH & WHERE TO FIND THEM

Fish have dominated the waters of Earth for hundreds of millions of years.

But what is a fish? Well, one thing is for sure: there are an awful lot of fish out there; more species than all the birds, reptiles and mammals put together! Four billion years ago, the whole planet was covered in *water*. Everything that lives in air today once crawled out from that water – even us humans. You might start to wonder... *are we actually a sort of fish too*? A bit confused? Well, don't worry. So am I!

Thankfully, scientists agree that to be considered a fish you need *gills* (we don't), you have to live in the *water* (not us), and you've most likely got *scales* and *fins* (definitely not us).

As incredibly different as some fish species might appear from each other, once you start looking at only the features and characteristics that they share, you soon discover that they have way more things *in common* with other fish than any other animal on Earth. In this section, we'll look at some of the different species of fish you might come across and where exactly you might find them.

MICRO FISH

Catching the very smallest fish in the water can be incredibly exciting...

They are very fast, very beautiful and very skilled at avoiding you!

I've already written about how you might use a net or bottle trap to snare these critters. Here are a few of the little fish you could be lucky enough to catch – each one full of colour and character.

Three-spined stickleback

Found in Europe and North America. Run your net along the edges, or leave a bottle trap in the margins of a pond, slow river or canal to catch this aggressive little fish.

Minnow

'Minnow' is a common name for many different fish species. Use a net or a bottle trap in the shallow waters of a clear-running stream.

Loach

There are more than 1,200 types of loach across Europe, Africa and Asia. Use a net in fast streams, turning over rocks and scooping into the space you create, or dip in the bed of a pond or river.

Bullhead or Miller's thumb

Native to Europe. Get your net, turn over the rocks and pebbles in a fast-flowing or clear stream, then scoop into the spaces underneath and look out for that great fat head!

Gudgeon

A monster among micro fish! Take out your whip or float rod in a European canal and chances are the mighty gudgeon will be the first fish you'll meet.

BIGGER FISH

There are some larger fish out there too...

First, find out just how big your likely target species could grow - some might be suitable on the lighter rod and line float-kit introduced in these pages, but others might need specialist gear and help from an experienced angler.

Pike

This ancient, notorious fish is perfectly built for sudden bursts of speed and has fang-like teeth along its jaws, earning it the nickname, the 'water wolf'. Hunt for these predators in freshwater lakes, ponds, slow rivers and canals.

CAUTION: If you have never been pike fishing before, then you must go with someone who knows how to safely set up your pike rig and has experience with unhooking and handling these very delicate fish.

Bass

There are several 'bass' species targeted by anglers around the world. All bass share similar shapes: spiky dorsal fins with large mouths compared to their bodies, perfect for swallowing their small fish prey. But despite their name and appearance, they are all from separate families of fish.

Catfish

Catfish are one of the most widespread and numerous fish on Earth, with more than 3,000 species, Most have muscular bodies and long whiskers to help them feel for food in the dark.

CAUTION: Catfish often grow very large and are powerful rod-breaking, line-snapping fish. Please fish with someone experienced and use suitably strong gear!

Eel

Tiny baby European eels make an extraordinary 3,000-mile, three-year journey, from the Sargasso Sea to the European continent. Here, adult eels can live for as long as eighty years before they reverse that journey, right back to the Sargasso, to breed and die.

North American eels also use the Sargasso for mating, while New Zealand's longfin eels breed in the sea near Tonga and can live more than a hundred years.

In Europe, the eel is an endangered species with a 95 per cent decline in numbers. If you catch one anywhere, always take great care of it.

Eels are very slippery to hold. If you run a fingertip down their body, you'll find you overload their sensory organs and they fall, briefly and harmlessly, into a trance, making them much easier to handle.

King Carp

The king carp species includes the 'common', which is fully scaled with golden scales completely covering its body and is closer to the look of the original carp species that first arrived in Europe from Asia around 700 years ago.

Perch

The perch is a beautifully camouflaged predator-in-miniature, and is found in rivers and lakes in Europe and North America. A small perch is a dream fish for a beginner angler, but watch out for its needle-sharp spines as you unhook it.

FLY-CAUGHT FISH

Most fly-caught fish belong to the salmonid family.

Take care: river fishing seasons for these fish are different to the dates for coarse fishing. Check before you cast out!

Grayling
This fish needs really clean, fast-running water. One of the truly great things about the grayling is that it will still feed on the very coldest days of winter, when all other species have lost their appetite.

Unhook a grayling in the river and place it with its head facing upstream to let it recover a while before releasing it.

Trout
Trout are found in rivers and lakes on every continent except Antarctica, and are a hugely popular fishing species.

Cast dry or wet flies into fast-flowing streams and rivers, taking care to use ones that look like what the trout might be eating at that time of day.

Salmon

To many, the salmon is the undisputed king of the fish.

Even though adult salmon don't feed in freshwater, they will snatch at flies and spinners that get in their way, perhaps out of instinct or maybe just to show who is boss! If you do hook one, be ready for a very strong fight.

Once the salmon is landed, give it plenty of time to rest and recover, with its head facing upstream, before releasing it to continue on its extraordinary journey.

GAME FISH

In the UK, most fly-caught fish are known as game fish, while other freshwater fish are referred to as coarse fish. The reason is a bit daft.

About 120 years ago, the lofty gentlemen catching fish like salmon or sea trout on the fly decided they wanted to make a distinction between themselves and other anglers who used bait such as worms and maggots.

From that point on, a fish like a salmon was known as a 'game' fish and everything else was a 'coarse' fish – another word to mean 'rough' – to make the gentleman's sport seem superior. Today, those 'coarse' and 'game' titles have stuck, but the original meanings are mostly forgotten.

SLOW-FLOWING WATERWAYS

Where are the fish lurking here?

The edges of the **riverbank** and shallow areas of water can hide fish.

Bridges and overhanging branches can create *shelter* for fish.

Fish like the quieter *slack* water in wider areas where boats can turn.

Look for the *inside step*, (where the river depth changes) – it traps food and attracts fish.

In the colder months, fish will shoal in the *deep water* of the central channel.

FLOWING RIVERS

Where can you catch a fish in a river?

The *'crease'* is where the fast flow meets the slow. Fish will hang in the slow and snatch food items as they pass by.

A *swirling eddy* behind a rock could shelter a salmon.

Barbel can feed on the *gravel beds* in slacker water where holes have been carved out by the flow.

Hidden in the tangle of tree roots might be a shoal of feeding chub.

Waving beds of weed can be full of fish.

Minnows, bullheads and loach might shoal in the *shallower waters* on the outside of a bend.

WELCOME TO THE CLUB

Honestly, the water really is magic!

Welcome to the great club of anglers. There are millions of us: all spread out across the world's waters; all casting our hopes after some fish of our *dreams*; all deeply in love with this slightly crazy hobby. Yet, beyond the fish we hope to catch, we are all probably fishing for our own secret reasons too.

When I am out on those banks, I forget about everything that might be giving me any trouble back home. I pour my worries into the water because my fishing is so all-consuming that I can't actually think about much else. Better still, I find that being outside in the fresh air, surrounded by *nature*, makes the things that worried me most not matter half as much as I thought they did.

But part of being an angler means taking *responsibility* for both the fish and the places where they live. When it really is time to reel up and head home, make sure you leave absolutely no trace of you ever having been there. Pick up all your litter, check for discarded line or any hooks that might be lying around.

Clean *freshwater* is the most precious thing. It sustains life as we know it and yet there is no other ecosystem under greater threat today. Since 1970, more than three-quarters of all the freshwater animals on Earth have vanished.

Pollution comes in many forms, but if your waterway smells like a toilet or has a *chemical* odour like a bad science experiment, and especially if you can see dead animals and fish, it is your duty as an angler to do something about it immediately. Tell the fishing club and call your environmental regulator. They will have an *emergency pollution hotline* and should send an officer to your location to test the water. Also, if you have a camera, take as many photos as you can.

There are other things you can do too. Volunteer to help clean up your waterway on organized *litter picks*. Become a member of the *Angling Trust* – it represents anglers and protects waterways across the United Kingdom – and, whether you live in the UK or not, follow the work of the *World Fish Migration Foundation*, as it continues its fight to free rivers from all the human-made obstacles to the great migratory fish (like eels and salmon) that are trying to complete their life's journey.

Our responsibility to water runs far deeper than it merely being a place where we can catch fish – being able to go fishing is a deep *privilege* and not a right.

GLOSSARY

Angler – a person who goes fishing.

Bait – food placed on a hook to attract fish to bite it and be caught.

Boilie – a type of ball-shaped bait made using foodstuffs that are attractive to fish.

Bottom-dweller – a fish that lives and feeds on the bed of a river, pond or lake.

Casting – the technique of launching a fishing line into the water.

Disgorger – a tool used to remove a fish hook from the mouth of a small fish.

Feeder fishing – fishing on the bed of a river, lake or pond using weighted bait.

Float – a buoyant piece of plastic attached to a fishing line that shows where the bait is located in the water and indicates when you have a bite.

Float fishing – a form of angling in which you suspend bait midwater under a float.

Fly – an artificial flying insect made from fur and feathers for use as bait.

Fly fishing – a form of angling using a non-edible 'fly' to lure fish onto a hook.

Forceps – a tool used to remove a fish hook from the mouth of a large fish.

Freelining – fishing with a rod that features just a hook, line and bait.

Groundbait – bait that is thrown into the water to attract fish to a specific area.

Hook – a bent piece of metal alloy at the end of a fishing line used to catch fish.

Knot – a type of loop used for fastening and tying fishing line.

Legering – see Feeder fishing.

Lure – a type of bait for catching fish.

Maggot – the legless larva of a fly or other insect that is a popular type of live bait.

Noodling – fishing for catfish with your bare hands.

Plummet – a weighted device for measuring the depth of a river, lake or pond.

Reel – a mechanism that stores and releases the length of fishing line on a rod.

Rig – an arrangement of fishing tackle that includes the hook, float and weights.

Rod – a long, thin piece of wood or synthetic material to which a fishing line is attached.

Ship – a swinging or sliding motion with which you move your rig out over the water

Spinner – a type of fishing lure.

Split shot – a small ball-shaped weight with a groove through its middle that can be attached to fishing lines.

Tackle – all the equipment needed for catching fish.

Trap – an item used to catch fish without a person being around, e.g. a basket, pot or bottle.

Waggler – a type of float attached by a line to a rod.

Whip – a type of fishing pole with a bendy tip.

Will Millard is a writer, television presenter, and life-long angler. Born and brought up in the Fens, Will lived and fished with remote Indigenous communities of the South Pacific in his award-winning BBC series *Hunters of the South Seas* and *My Year With the Tribe*. He has presented several other BBC documentaries on rivers and fishing, and regularly writes for *Fallon's Angler* magazine, *BBC News*, *Countryfile*, and the *Guardian*, among many others.

Will is an ambassador of World Fish Migration Day, the Angling Trust, and Keep Wales Tidy, a Fellow of the Royal Geographical Society and the proud patron of Mobile Education Partnerships, who provide teacher training and resources for schools in Myanmar.

When he isn't out exploring, he's either writing books at his kitchen table or fishing the rivers and lakes around his South Wales home.

My First Day Fishing © 2024 Lucky Cat Publishing Ltd
Text © 2024 Will Millard
Illustrations pages 12-17, 28-29, 34-35, 46-49, 56-59, 72-73 © 2024 Nick Hayes
Illustrations pages 2-11, 18-27, 30-33, 36-45, 50-55, 60-71, 74-92 © 2024 Joanna Lisowiec

First Published in 2024 by Magic Cat Publishing, an imprint of Lucky Cat Publishing Ltd,
Unit 2, Empress Works, 24 Grove Passage, London E2 9FQ, UK

The right of Will Millard to be identified as the author of this work and Nick Hayes and Joanna Lisowiec to be identified as the illustrators of this work has been asserted by them in accordance with the Copyright, Designs and Patents Act, 1988 (UK).

No part of this publication may be reproduced, stored in a retrieval system, or transmitted, in any form, or by any means, electrical, mechanical, photocopying, recording or otherwise without the prior written permission of the publisher or a licence permitting restricted copying.

A catalogue record for this book is available from the British Library.

ISBN 978-1-915569-28-8

The illustrations were created digitally
Set in Edith, Modern Love and Noyh A Hand

Published by Rachel Williams and Jenny Broom
Designed by Sophie Gordon

Manufactured in China

9 8 7 6 5 4 3 2 1